APPLE

APPLE

ODYSSEYS

PAMELA DELL AND SARA GILBERT

CREATIVE EDUCATION · CREATIVE PAPERBACKS

Published by Creative Education and Creative Paperbacks
P.O. Box 227, Mankato, Minnesota 56002
Creative Education and Creative Paperbacks
are imprints of The Creative Company
www.thecreativecompany.us

Design by Graham Morgan
Art direction by Tom Morgan
Edited by Jill Kalz

Images by Alamy (Mirrorpix, Richard Levine), Getty (Apple/Getty Images, Focus On Sport, Gilles Mingasson, Justin Sullivan, Paul Harris, Science & Society Picture Library, Tayfun Coskun/Anadolu Agency), Pexels (Tuur Tisseghem, Zetong Li), Unsplash (Alexander Shatov, Andres Urena, Avi Richards, Brett Jordan, Ethan Rougon, Jimmy Jin, Medhat Dawoud, Museums Victoria, Phil Aicken, Solen Feyissa, surasak_ch), Wikimedia Commons (Ben Stanfield, Bernard Gotfryd, Binarysequence, Felix Winkelnkemper, Gregory Varnum, MIKI Yoshihito)

Library of Congress Cataloging-in-Publication Data
Names: Dell, Pamela, author. | Gilbert, Sara, author.
Title: Apple / by Pamela Dell and Sara Gilbert.
Description: Mankato, Minnesota : Creative Education and Creative Paperbacks, [2025] | Series: Odysseys in business | Includes bibliographical references and index. | Audience: Ages 12–15 | Audience: Grades 7–9 | Summary: "A business survey for young adults of the computer and electronics giant Apple, covering the global company's products and technology, history, and founders, Steve Jobs and Steve Wozniak. Includes sidebars, a glossary, and further resources"— Provided by publisher.
Identifiers: LCCN 2023046721 (print) | LCCN 2023046722 (ebook) | ISBN 9781640269156 (library binding) | ISBN 9781682774656 (paperback) | ISBN 9798889890836 (ebook)
Subjects: LCSH: Jobs, Steve, 1955–2011—Juvenile literature. | Wozniak, Steve, 1950-—Juvenile literature. | Businesspeople—United States—Juvenile literature. | Apple Computer, Inc.—History—Juvenile literature. | Computer engineers—United States—Biography—Juvenile literature.
Classification: LCC QA76.2.J63 D45 2024 (print) | LCC QA76.2.J63 (ebook) | DDC 621.39/0922—dc23/eng/20231116
LC record available at https://lccn.loc.gov/2023046721
LC ebook record available at https://lccn.loc.gov/2023046722

Printed in China

Apple cofounder Steve Jobs

Apple cofounder Steve Jobs and the first Macintosh personal computer, 1984

CONTENTS

Introduction

In January 1984, a TV commercial for Apple Computer, Inc., aired during Super Bowl XVIII. The ad, cast in dull blue-gray tones, echoed George Orwell's dystopian sci-fi novel *1984*. It opened with a horde of robot-like men stomping into a hall. There, they mindlessly watched their leader, like the novel's Big Brother character, rant from a black-and-white movie screen.

OPPOSITE: The dark, unsettling commercial teasing Apple's Macintosh in 1984 aired only once.

Amidst this sea of cloned conformity, a sledgehammer-wielding woman ran into the hall. Dressed in orange shorts and a white tank top, she was the single splash of color in the scene. "We shall prevail!" Big Brother declared just as the woman hurled her sledgehammer at the screen, which exploded in a burst of light. Two sentences appeared, backed by a voiceover:

> *On January 24th, Apple will introduce Macintosh. And you'll see why 1984 won't be like "1984."*

Orwell's novel about conformity, oppression, and the elimination of free thought was in direct contrast with Apple's mission.

OPPOSITE Customers enter Apple Fifth Avenue—a 24/7 retail store in New York City—through a giant glass cube.

That ad signaled things to come for Apple, a company officially **incorporated** in 1977. Its loud-and-clear message was this: Something's about to shake up the **status quo**.

In its nearly half-century of existence, Apple has had epic fails and spectacular successes. But that ad hinted at the force for change that the company would come to represent. Taking visionary risks has paid off. By April 2023, Apple was the most **profitable** company in the world.

Apple Seeds

Just as a fruitful apple tree sprouts from a tiny seed, so Apple, Inc., grew from a seedling idea that sprouted in the fertile soil of two young minds. It was the summer of 1971 in Cupertino, California. A mutual friend introduced 21-year-old Steve Wozniak to 16-year-old Steve Jobs.

OPPOSITE: Built in 1976, the Apple-1 personal computer could be made portable by carrying its parts in a briefcase.

"Woz" was a college dropout. Jobs was still in high school. But the two Steves had a shared interest in electronics. At the time they met, Cupertino was part of the newly forming "Silicon Valley," a region just south of San Francisco. The area was rapidly becoming a magnet for the tech industry. Big, successful companies such as IBM and Xerox had landed there. Interest in small, personal computers (PCs) was on the rise, too. Woz and Jobs had come together in the right place at the right time.

By 1973, both young men had found jobs in their chosen industry. Woz worked full-time designing handheld calculators for Hewlett-Packard. Jobs, who had dropped out of Reed College in Oregon, was a technician for Atari, then the world's largest video game company. In 1975, the friends joined a local techie users group called the Homebrew Computer Club. Activities there inspired them to try building a small computer that would appeal to the everyday consumer. Woz, the duo's computer designer and builder, got to work. The result was a simple machine based on a $20 microprocessor—the component that allows the computer to respond to commands and powers its operations.

Jobs recognized the potential of what they had and suggested they try to sell a few. But Woz had a strict contract with his employer. He needed a legal release from Hewlett-Packard to produce and promote his own electronics.

After deliberating, both he and Jobs left the companies they worked for and, on April 1, 1976, formed their own company. Jobs came up with the name Apple Computer. It was a nod to time he had spent working on an apple farm in Oregon. Jobs later said he liked the name because it "sounded fun, spirited, and not intimidating."

Looking for customers for their first product, Jobs approached Homebrew member Paul Terrell. Terrell owned a local store that sold computer kits, parts, and accessories, mainly to hobbyists. Jobs asked Terrell for an order and got one on the spot—50 units that

Large technology company campuses started springing up in Silicon Valley in the 1970s.

1984

In 2023, a 30-second Super Bowl ad cost $7 million. Thirty-nine years earlier, Apple had paid just $1 million for its twice-as-long 60-second Super Bowl ad spot. The ad was directed by the legendary Ridley Scott, whose cult classics include films such as *Alien* and *Gladiator*. Apple's **board of directors** hated the ad, but the network wouldn't refund the money. So, "1984" got its chance—and struck a chord with viewers everywhere. That new-concept storytelling ad won multiple advertising awards for "best commercial of all time." And it started a trend. Viewers now expect to see advertisers' very best ads during the Super Bowl.

Terrell agreed to buy for $500 each. In July, two months after Apple "went live," that first sale was delivered—for $25,000 cash. The product, called the Apple-1, was essentially a simple **circuit board** seated in a wooden frame, a build-it-yourself kit. Users needed to add their own display, keyboard, and power supply. The Apple-1 was the first computer to allow input and output using monitor and keyboard—even if buyers did have to add those parts themselves.

In working to meet a tight delivery date, Woz and Jobs had brought in a third partner named Ronald Wayne. At the time Terrell had placed his order, the

Apple team had taken a $5,000 loan and gotten a line of credit to buy parts and supplies. Now they were able to pay off those debts, approximately $17,000 worth. The remaining $8,000 was Apple's first profit.

The two Steves were optimistic about their new startup. Wayne was less so. Worried about financial stability, he decided to leave. Thanks to Jobs's sales and marketing skills, however, Apple kept landing orders, some from as far away as the Midwest and New York. Woz priced the Apple-1 at $666.66 each—more than $3,500 in today's market.

When he had received his order, Terrell had made a wise suggestion: Why didn't the Steve-team build and sell fully assembled models themselves? A year later, Wozniak did just that. His new model, the Apple II, hit the market on April 15, 1977. Woz had singlehandedly designed its circuit boards, operating system, and the computer's hardware and plunked it all into a light-colored, modern-looking case.

Unlike the Apple-1 and reflecting Terrell's advice, the Apple II needed no assembly. Even better, it had high-resolution graphics capable of displaying up to six onscreen colors. To acknowledge this cutting-edge advancement, Jobs came up with the now-famous rainbow-striped Apple logo. Like every other computer of the day, the Apple-1 displayed only black and white.

The “bite” in the Apple logo was meant to keep people from mistaking the apple for a cherry.

Once people saw the Apple II at computer trade shows, the Apple-1 was all but forgotten. The user-friendly personal computer was about to get real.

Producing the Apple II had required even more money than the Apple-1. In 1977, Jobs and Woz found an angel investor named Mike Markkula. A retired 35-year-old millionaire, Markkula offered them $250,000 and joined the company as its third employee. After Woz rented a small headquarters space in Cupertino, Apple officially incorporated on January 3, 1977. A month later,

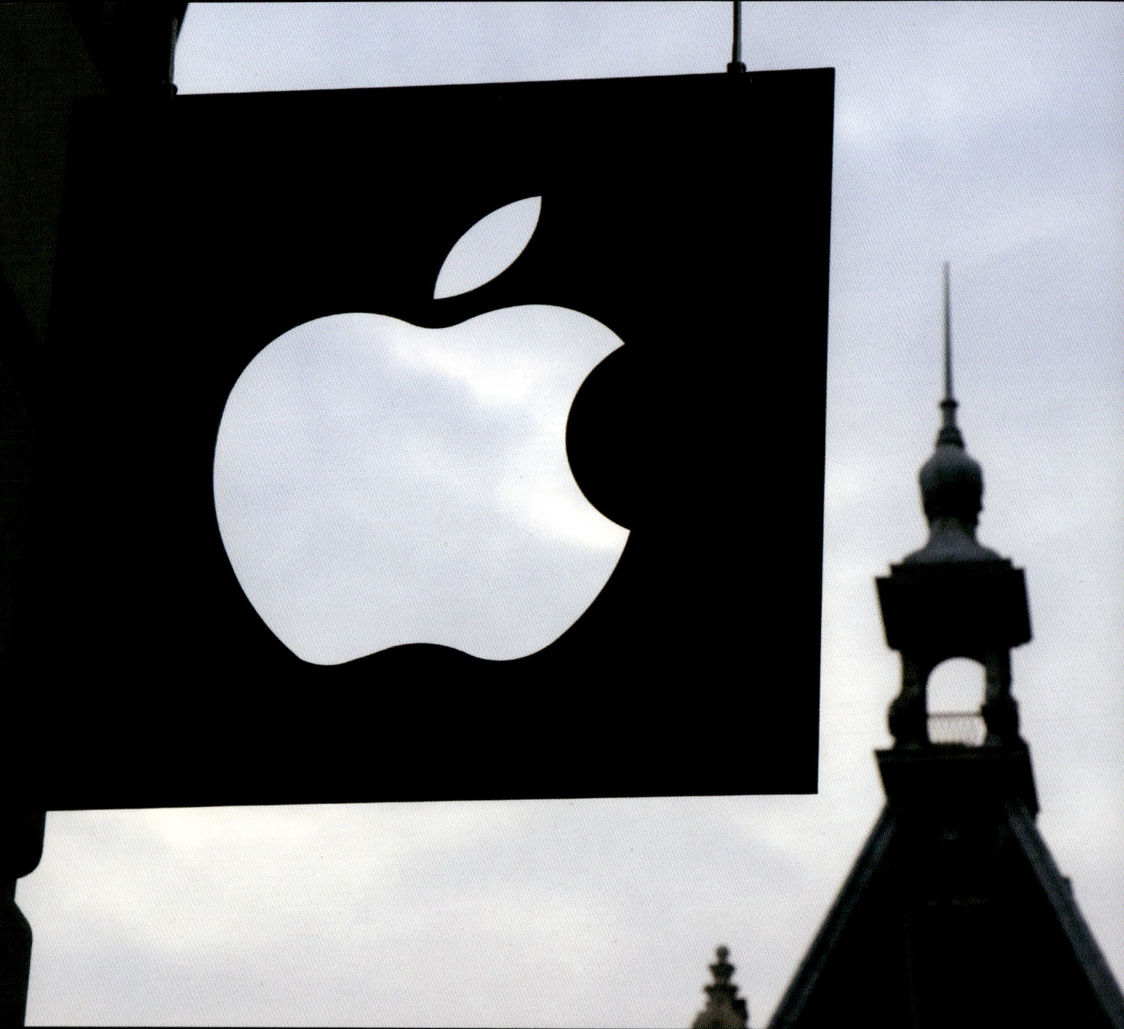

Logo Evolution

Apple's logo has been through numerous changes. It began as an illustration of 17th-century scientist Isaac Newton sitting under an apple tree along with the words "Apple Computer Co." In 1977, the apple with a bite out of it appeared, with colorful stripes representing the Apple II's color capabilities. From there, that apple changed its colors many times. It went from translucent blue (1998) to black (1998) to chrome (2001). Then there was the "Sliced Apple" (2007) and, finally, a dark gray (2017) apple. Now the logo is so well-recognized that the word "Apple" isn't even

Markkula hired his friend Michael Scott as the company's first chief executive officer (CEO). Markkula believed the two young founders needed someone with more business experience to guide them. Now, Apple Computer, Inc., was a real, fully operational business.

The company quickly began to grow. With its attractive casing and standard keyboard, the Apple II sold well at a retail price of $1,298. And by 1980, Apple had about 1,500 employees. New computers were also in the pipeline: the Apple III, the Lisa, and a little machine they'd named the Macintosh, or "Mac."

With Apple's continuing success, the founders felt sure enough to hold an initial public offering. On December 12, 1980, the company offered 4.6 million shares of stock for public purchase. The shares sold out in minutes (at $22 each), and by the end of the day, Jobs, Woz, and Markkula were multimillionaires. Apple itself was suddenly worth $1.8 billion.

The year 1981 brought changes. Scott left the company after abruptly terminating many employees. Markkula temporarily took over as CEO. Jobs, who had been overseeing the Macintosh development group, became chairman of the board. That same year, Wozniak took time off to recuperate after surviving a small plane crash.

Then, in 1983, Jobs lured John Sculley away from the presidency of PepsiCo to become Apple's new CEO.

Under Sculley's watch, the affordable Apple II continued to sell. The famous—or infamous—1984 Super Bowl ad had brought the groundbreaking Mac to public attention, too, with sales following. Apple was on an upswing.

Shortly after John Sculley (*center*) joined Apple's leadership team in 1983, cofounders Jobs (*left*) and Woz (*right*) left the company.

A4
LaserWriter and
LaserWriter Plus

Missteps and Losses

During John Sculley's 10 years as CEO at Apple, the Mac and the Apple II remained the company's bestsellers. With its graphical user interface (GUI) and other features, the Mac eventually outdistanced the popular Apple II. Computer sales grew from $569 million to $8.3 billion. Sculley also brought the world easy and accessible **desktop publishing** by hooking up Aldus Pagemaker software with the Apple LaserWriter printer.

OPPOSITE: The LaserWriter printer debuted in 1985 at a cost just under $7,000 (about $19,000 in today's dollars).

In 1991, Apple introduced the PowerBook, which raked in another $1 billion in sales just in that year alone.

The PowerBook was a trimmer, more powerful version of the company's earlier attempt at a portable computer, which had utterly failed. After that flop, Sculley's leadership brought on further fiascos. The Lisa computer tanked badly. Sculley's own pet project, a personal digital assistant (PDA) called the Newton, was ahead of its time and went nowhere. Other projects didn't pan out, either.

In the mid-1980s, Sculley also signed a self-destructive deal with Microsoft CEO Bill Gates. Through that deal, Microsoft—Apple's biggest competitor—ultimately gained the right to create its own GUI, which led to the development of Microsoft Windows. This, in turn, allowed Microsoft to dominate the personal computer industry, a horrendous outcome for Apple and its profit margins.

The hiring of Sculley also ended up backfiring big-time on Steve Jobs himself. The two men had meant to run the company together, but their constant conflict cost them both. By 1985, Sculley felt Jobs was butting in where he wasn't wanted. In what has been called a "boardroom brawl," Sculley forced Jobs out of the company entirely. Wozniak, who had returned

Singer-songwriters Yoko Ono and John Lennon

Think Different

In 1997, Apple was on the brink of bankruptcy. So, Steve Jobs approved the unique "Think Different" ad campaign to refresh Apple's image. A long list of exceptional people (such as physicist Albert Einstein and civil rights protester Rosa Parks) appeared in ads accompanied by only the Apple logo and the words "Think different." The ads appeared in magazines and on billboards and buses. On TV, the commercial included spoken words, in part: "Here's to the crazy ones . . . The rebels. The troublemakers . . . The ones who think differently . . . [T]he people who are crazy enough to think they can change the world are the ones who do."

to Apple in 1982, also walked away in 1985. Apple was without its two primary founders.

By 1993, the writing was on the wall for Sculley. In the fourth **quarter** of 1992, the company had shown profits of $97.6 million. By the same quarter of 1993, Apple was down deep in the profit pits, making only about 3 percent of the year before—a mere $4.7 million. On October 15, 1993, the day after the dismal profit report came out, Sculley "left the building" forever. Apple was at its lowest point.

Next up as Apple CEO was a man named Michael Spindler. "The Diesel," so-called because of his 6-foot-4-inch (1.9-meter) frame and his intense working style, wasted no time changing things up. Despite its bad bottom line in 1993, Apple was still the world's most profitable computer company. But The Diesel wanted to do better. He slashed the research and development budget. He put an end to several ongoing projects. And he cut 2,500 employees.

These were extreme moves, but they paid off. Over the next four quarters, Apple saw strong and steady growth. So strong, in fact, that Spindler declined what would have been some potentially high-profit **merger** deals. He saw higher profits in Apple's future and wanted to keep the company independent.

Those profits started to pour in with the 1994 release of a souped-up, speedy computer aimed at the business community. In only 10 months, that computer, the Power Mac, sold one million units. But Spindler, CEO number four, also had trouble on his hands. As well as other serious problems, Apple's top guns hadn't anticipated the high demand among consumers. Production just couldn't keep up, quickly resulting in a huge backlog of orders.

Spindler soon got the axe, too, and Apple board member Gilbert Amelio took over in February 1996. Amelio was expected to turn the company around. But he never lived up to his promise. Apple's prospects for staying afloat had been declining since 1990, and 1997 put a cap on that fact: In only six months, the company had suffered losses of $740 million, and 2,800 employees had been laid off. It wasn't the best way to celebrate Apple's 20th anniversary.

With Amelio at the helm, Apple suffered enormous losses—a shortage of $1 billon, to be exact. The total loss was so

Mighty Mac

The groundbreaking Macintosh computer had several exceptional and never-before-seen features. Its GUI meant that instead of having to type computer commands in text, users had a visual, intuitive way of working. All taken for granted today, these included a hand-directed input device called a "mouse"; onscreen frames called "windows," which could be shuffled around and overlapped like pieces of real paper; clickable pictures or "icons," such as a tiny trash can for getting rid of documents; and copy, cut, and paste functions. The Mac "desktop" functioned in a way that was quickly familiar to anyone with a real desktop.

A 3D model of a 1984 Macintosh

extreme that it basically cancelled out all the profits Apple had earned since 1991. The board felt Amelio wasn't dynamic or farsighted enough to run the company they envisioned. In July 1997, exactly 500 days after he started, Amelio, too, was gone.

But Amelio had done a foresighted thing before he'd been asked to leave. Trying to stop the bleed at Apple, he'd bought NeXT, the tech company Steve Jobs had started, in February 1997. Amelio also managed to get Jobs back as a consultant. By then the Apple founder had 12 years of business experience. This gave him new authority and insight he hadn't had when he'd left.

Now, with Amelio out, Jobs agreed to act as Apple's **interim** CEO until the board of directors found a replacement. That move would signal one of the biggest business turnarounds ever, an almost meteoric rise out of near-death. Jobs took charge, ousting the board, including Markkula.

He also changed the nearly bankrupt company's focus. He turned Apple Computer, Inc., back to producing computers almost exclusively: two desktop models and two portables.

In another big, bold move, Jobs engineered a partnership deal with Microsoft. Mac loyalists were disappointed, but the deal benefitted both companies. "Apple has to move beyond the point of view that for Apple to win, Microsoft has to lose," Jobs declared. All in all, the deal meant money in the bank for Apple. By the first quarter of 1998, the company was again making profits.

With Jobs back in charge, Apple fans began to feel a sense of optimism. Maybe the company would survive. Little did anyone know how far that simple hope would be surpassed. Behind the scenes, Jobs had a secret weapon—one that would change the world of information technology beyond anyone's wildest imagination.

Finder
Ablage
Bearbeiten
Darstellung
Gehe zu
Fenster
Hilfe
Do., 01:03
Programme
Zurück
Vor
Darstellung
Computer
Privat
Favoriten
Programme
Suchen
Number One1
1 von 28 Objekte ausgewählt, 6,6 GB verfügbar
Benutzer
Firefox 2.0.0.2.dmg
Library
Programme
System
Temporary Items
Acrobat Reader 5.0
Adobe Reader 7.0
Adressbuch
AppleScript
Chess
Dienstprogramme
Digitale Bilder
EPSON Printer Utility
Firefox
iCal
iChat
iMovie
Internet Explorer
Internet-Verbindung
iPhoto
iTunes
Mail
Name iMovie
Art Programm
Größe 28,6 MB
Erstellt 14.07.2002
Geändert 15.01.2007
Version 2.1.2
iMac

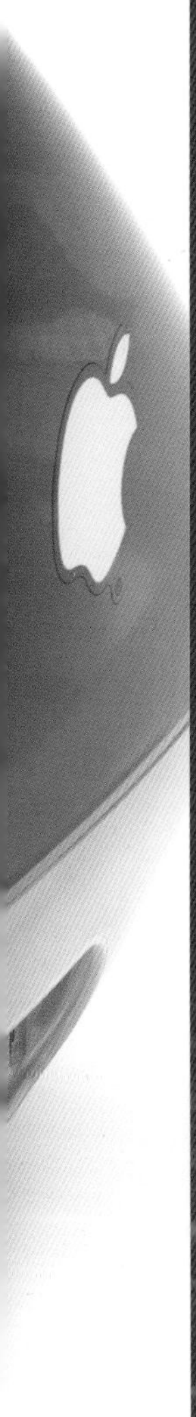

i Is for Apple

When it came to Apple, Steve Jobs had always been a perfectionist. He insisted on delivering a beautiful, uniquely functional product that people loved. The billions he was worth meant little to him compared to his drive to give the world something of value.

OPPOSITE: The letter "i" in many Apple product names has come to represent words other than *Internet*, such as *individual*, *instruct*, *inform*, and *inspire*.

In this, Jobs met a kindred soul. Jony Ive was a British-born industrial designer who had joined Apple in 1992 at only 25 years old. A true design **prodigy**, Ive was, like Jobs, a detail-minded perfectionist. Both valued a nonconformist approach to things, and Jobs was soon speaking of Ive as the second highest-ranking person at Apple, after himself. Now 31 years old and appointed Apple's chief design officer, Ive began spending countless hours with Jobs, conjuring up ideas for new and **innovative** technology.

The first indication of what this collaboration meant came in August 1998, when the iMac G3 went on sale. It was a computer unlike any other. It wasn't a square, boxy machine. It wasn't a dull black, beige, or gray color. The radical new design, the first to be credited to Ive,

Jony Ive, Apple's former chief design officer, in 2004

iPhone Sensation

With its release in 2007, iPhone started a revolution in the world of tech. Every year since 2012, Apple has introduced at least one new model. These new versions generally include upgraded features—and higher prices. The sleek 2023 iPhone 15 came in four models, with the same size options that iPhone 14s had. The display on the two larger "Pro" versions, with all Apple's best features, was bigger than 6 inches (15 centimeters). Not surprisingly, iPhone is the best-selling smartphone on the planet. In 2022 alone, iPhone sales totaled more than half of Apple's entire annual revenue.

was a white-trimmed, bright translucent "Bondi blue," giving users a peek at its innards. Its shape was what some people called a gumdrop. Its appeal was instantaneous.

Apple hyped the iMac as built for the Internet age—thus the small "i" added to the front end of the computer's name and other products thereafter. Later iMac models came in a range of fruit-candy colors, or "flavors," such as strawberry and tangerine. Eventually, that range expanded to 13 colors total. The iMac G3 had many tech improvements as well. It was loaded with memory and power at a reasonable price ($1,299). Most desirable was its internal modem for instant Internet access, something other computers lacked. With 150,000 pre-orders, Apple's stock price skyrocketed. One journalist wrote that the iMac G3 had "cemented Jobs's legacy as a forward-thinking tech visionary."

Jobs and Ive never looked back. In January 2000, Jobs announced he was signing on permanently as Apple CEO. Thanks to Ive and his design dream team, Jobs began releasing new eye-catching products and updates every year, sometimes even more frequently. Not much later, Apple's ideas for new technology began to expand well beyond computers.

This led Jobs to announce, at the January 2007 MacWorld Expo, that Apple Computer, Inc., was changing its name to simply Apple, Inc. This slightly new name signaled the brand's expansion into consumer electronics and products other than just computers.

The decision made sense. In the dozen years after the iMac G3 made its debut, Apple had numerous big non-computer knockouts, including:

January 2001: Apple iTunes – A massive hit, this product was second only to Apple's operating system software in terms of importance for the company's success.

May 2001: Apple retail store – Apple's first two brick-and-mortar stores opened in Glendale, California, and

Print ads for the iMac G3 described the colorful computer as "Chic. Not geek."

McLean, Virginia. By 2024, the retail chain had more than 520 stores throughout 26 countries, 273 of them in the United States.

October 2001: Apple iPod – The small, sleek MP3 player, a Jony Ive-designed device for portable music, could store up to 1,000 songs. Less than five years later, 100 million iPods had been sold in 11 different versions. The product was discontinued in 2022.

January 2007: Apple iPhone – The biggest advance in cell phones to date, this smartphone included a touch screen, Internet capability, and iPod-like music storage. Before the end of summer, more than four million people around the world owned one.

January 2010: Apple iPad – At only $499, the light, thin, touch screen tablet was designed for Internet browsing, sending email, reading digital books, and playing music, movies, and games.

On October 5, 2011, bad news shook the world. Steve Jobs, the man who had become the **iconic** symbol of Apple itself, had died, at just 56 years old. Apple's stock initially plunged but then began to improve later that day.

A little less than a year later, on September 12, 2012, Apple released its latest smartphone, the iPhone 5. Even with several flashy new features, it didn't get a particularly spectacular reception. One product reviewer commented, "The iPhone 5 is completely amazing and utterly boring." Nevertheless, the iPhone 5 marked up sales of more than $125 million.

Although the iPhone was a hit from the start, there was nervous speculation about Apple's future after Jobs's death. Would Apple lose its "mojo" with him gone? Some

The iPod couldn't compete with the iPhone's multimedia capabilities and began phasing out in 2014.

Smartwatch History

Smartwatches aren't an entirely 21st-century thing. Various "calculator watches" and "computer watches" have been around for nearly 100 years. The first attempt came in 1927. It was a wristwatch that had a scrolling map to help wearers plan their driving routes. From 1972 through the 1990s, many companies introduced their own models, all with relatively primitive technology. In the next millennium, numerous brands, including Sony, Nike, and Samsung, got into the game. Some models had sleek, cool designs, while others were clunky and gigantic. What they could do varied, too. Apple Watch was a real leap ahead.

pessimists believed the company was doomed. But in August 2011, when Jobs had resigned due to his failing health, he had appointed Tim Cook as CEO, and Cook had already begun proving himself in that role.

Cook was a low-key man who knew how to take care of business. A competent leader, he made major supply-chain improvements and had a more democratic managing style than Jobs had. Whereas Jobs had his hand in every aspect of the business, Cook gave freer rein for decision-making to high-level executives. Under his lead,

Apple would prove its ability to keep on keeping on, in the best possible sense.

Chief designer Ive took Jobs's death hard, but he continued to deliver. In 2012, Apple introduced its EarPods, an upgrade from the Earbuds that had been packaged with the iPod. With their signature white color, much improved sound, and a comfier shape for the ear, EarPods caught on like wildfire. It wasn't until 2016 that Apple released its **Bluetooth** version, the AirPods. Not surprisingly,

TAKEAWAY

Whereas Jobs had his hand in every aspect of the business, Cook gave freer rein for decision-making to high-level executives.

those were an instant hit, too. About three years later, AirPods 2 came along. AirPods remain the most popular wireless earbuds ever made.

Even with the success of AirPods, Ive was itching for something else with which to wow the world. His energies turned to wearable tech. That translated into Apple's next blazing triumph—a super-smart wristwatch.

Users can dress up or dress down their Apple Watches with rubber, metal, or leather bands.

Apple's first smartwatch was released in April 2015. Given their iPhone brand loyalty, consumers anxiously waited, expecting the best. And yet again, Apple delivered. Simply called Apple Watch (rather than "iWatch," as fans had expected), the device was beautifully designed. The Series 1 watch came in four models. It also had a touch screen, dozens of apps, health and fitness tracking abilities, and a never-before-seen interface.

As always, Apple showed up as leader of the pack. Apple Watch wasn't perfect, but it surpassed any other smartwatch on the market that year. Still, Apple has always been about perfection. Each year thereafter, the company released a new series of its product, always with additional new features and capabilities.

Future Growth

From the start, it seemed clear that the original Apple Watch was destined to join the long list of classy, trend-setting designs by Apple's chief design officer and his team. But Ive's crowning—and physically largest—achievement came in 2017. That was Apple Park, Apple's sprawling 176-acre (71-hectare) corporate headquarters in Cupertino, California.

OPPOSITE: Beauty and robust functionality have always gone hand-in-hand at Apple.

Apple Park's earliest inspiration had been Steve Jobs himself. Jobs had dreamed of one day giving Apple a headquarters unlike the usual, dull industrial center. He had wanted something that would look more like a peaceful nature preserve. And, of course, anything that carried the Apple name would require the brand's upscale look and feel. With these things in mind, Ive brought in the prestigious British architectural firm Foster + Partners. Just as he had with Jobs, Ive pored over every detail of the design (even down to the look of the elevator buttons) in collaboration with the firm's principal, Norman Foster.

The result—the park's crown jewel—was a stunning, ring-shaped main building. With a footprint greater than 2 million square feet (185,806 square meters), it houses more than 12,000 employees. This

An aerial shot of Apple Park

The Apple Park Visitor Center, flanked by olive trees, welcomes guests with its tall glass walls and blond wood accents.

humongous flying-saucer-like construction also has four-story ground-to-sky glass walls that reflect like a wall of fire in the sun. No one could miss it.

In a major nod to Jobs's original intention, 80 percent of the campus is green space. Also, unlike other tech-industry headquarters, Apple Park's design includes a lavish visitor center, with an Apple Store, a roof terrace and spacious café, and an exhibition hall. The visitor center is the only area open to the public. The rest is entirely private—the enviable, secretive employee workspace.

Inside the Apple Park Visitor Center

Apple Park opened to employees in April 2017. In some of the first aerial photos, the building really did look like a UFO hovering just above a woodsy landscape. The global reaction to the so-called spaceship was intense and mixed. Ive was known to refer to Apple Park as his home, and he didn't mince words when it came to the naysayers. "We didn't make Apple Park for other people," he told a reporter. "So, a lot of the criticisms are utterly bizarre because it wasn't made for you! And I know how we work, and you don't!"

In 2019, another blast of Apple news made waves in the tech universe. On June 27, the company announced that Jony Ive would be leaving to open his own design firm. It came as a surprise—even a shock—to Apple fans.

CEO Tim Cook said in his announcement, "Jony is a singular figure in the design world, and his role in Apple's revival cannot be overstated, from 1998's groundbreaking iMac [G3] to the iPhone and the unprecedented ambition of Apple Park."

The last product Ive officially signed off on was the 2019 Mac Pro. But he is also said to have had a hand in designing the high-class AirPods Max, released in late 2020. Still, Apple Park was a one-off, something that would never be repeated. As one writer stated, "Apple Park is going to be a 100-year testimony to Jony Ive's design."

Apple Watch Features

The Series 2 Apple Watch, introduced in 2016, included GPS. Also, for swimmers, it could log speed, number of lengths, and even keep track of the types of strokes a swimmer used. Series 3, the following year, allowed users to make and receive calls, stream music, and talk directly to the watch itself. This model also had more health features, such as an advanced heart rate monitor. In 2020, the first year of the COVID-19 pandemic, the Series 6 watch could keep track of the body's amount of oxygenated blood and included the CarKey, which could start a new car, if you had one.

Ive was gone, but his influence lived on. Headed by Tim Cook, the company continued to pull in mega-profits. In August 2018, Apple became the first U.S. company to be worth $1 trillion. Only two years later, in August 2020, that number doubled to $2 trillion. And the company hit the $3-trillion marker on January 3, 2022.

Since 2008, 52 percent of Apple's profits have come from the iPhone. The company's next biggest moneymaker is its services, which now include cloud storage, video games, fitness, and, of course, music and video streaming. These services brought in $53 billion in 2020 alone.

Jobs's extraordinary futuristic vision, Ive's ingenious product design, and the superior technological expertise that Apple products reflect have much to do with this.

But Cook has had the business savvy to keep the company evolving in multiple ways. As well as managing product releases, he has overseen Apple's expansion into markets such as India and China. He has formed important business alliances across the globe. Guided by Cook, Apple is committed to being a 100-percent renewable energy company and to fostering diversity and inclusion in the workplace. Apple also strives to be environmentally responsible. Those efforts range from rooftop solar panels (which supply the headquarters building with all of its energy) to chic-looking, Earth-friendly product packaging.

Apple, Inc., will forever be known as the superstar company whose ideas and products permanently changed the way people interact with technology. The brand has grabbed public imagination as no other tech

Apple advertising has always married bold images with simple, clear messaging.

company has. Historically, it has shaped and reshaped the entire landscape of computing, being almost singlehandedly responsible for the personal computer revolution.

Apple is clearly a company looking into the future. But what is that future exactly? It's difficult to predict the magic happening inside that secretive "spaceship." One strong possibility, though, is that in a few years, Apple will introduce its version of virtual reality technology.

What began as a small seedling—the work of two young college dropouts—has grown into a global megabrand with deep roots and many branches. Apple was and still is the company that, with its "rock star" leader Steve Jobs, revolutionized the use of media for an entire generation. Based on its dramatic past, more surprises will be in store . . . That rumored Apple electric automobile may not be so farfetched!

Worker Perks

A few of the perks, or benefits, Apple Park workers have access to include a pond, an orchard, and a meadow, all inside the building's circle. They also enjoy 2 miles (3.2 kilometers) of running and walking paths throughout the grounds. Ive and his colleagues designed the Apple Stage enclosed in the circle, too. The stage's rainbow-colored arch echoes Apple's former logo. The stage is used for concerts and other employee events.

Selected Bibliography

Christoph. "Timeline: The History of Apple Since 1976." Mac-History. December 16, 2022. https://www.mac-history.net/2022/12/16/timeline-the-history-of-apple-since-1976.

Isaacson, Walter. "The Real Leadership Lessons of Steve Jobs." *Harvard Business Review*. April 2012. https://hbr.org/2012/04/the-real-leadership-lessons-of-steve-jobs.

---. *Steve Jobs*. New York: Simon & Schuster, 2011.

Kahney, Leander. *Jony Ive: The Genius Behind Apple's Greatest Products*. New York: Portfolio/Penguin, 2013.

Lashinsky, Adam. "Inside Apple: How America's Most Admired—and Secretive—Company Really Works." *Business Plus*. October 15, 2012.

Podolny, Joel M., and Morten T. Hansen. "How Apple Is Organized for Innovation." *Harvard Business Review*. November-December 2022.

Simon, Michael. "Apple at 40: Remembering When Steve Jobs Went Electric." *Macworld*. April 4, 2016.

Glossary

angel investor someone who contributes money to a business, especially a startup company, for financial return

Bluetooth a standard for wireless connection between computers, mobile phones, and other electronic devices

board of directors a committee that governs the actions of a business or organization

circuit board a set of electrical connections that allows a device to perform specific functions

desktop publishing the production of printed material from a printer linked to a desktop computer that uses document-creating software

executive a decision-making leader of a company, such as the president or chief executive officer (CEO)

iconic widely known and easily recognized as representing something for its excellence

incorporate to legally form into a company

initial public offering the process of offering shares in a company to the public for the first time

innovative original or advanced beyond the usual; applying new methods of doing things

interim temporary, usually until a permanent replacement can be found

MacWorld Expo an annual information technology trade show whose purpose was to highlight Apple's Mac platform; the last MacWorld Expo was held in 2014

merger the combining of two or more entities into one through a purchase or combining of interests

prodigy someone, usually young, who has exceptional skills or abilities beyond what is typical

profitable having money left over once expenses are subtracted from revenues

quarter one of four three-month intervals that together comprise the financial year

status quo the existing state of affairs

stock shared ownership in a company by many people who buy shares, or portions, of stock, hoping the company will make a profit and the stock value will increase

Websites

The History of Apple in Photos

https://www.businessinsider.com/history-of-apple-in-photos-2015-8

Review a photo history of Apple.

How Jobs & Wozniak Used Cognitive Friction to Change the World

https://www.a.team/mission/all-time-a-teams-how-jobs-wozniak-harnessed-cognitive-friction

Learn about the role friction played in the foundation of Apple.

Steve Jobs—Complete Biography, History, and Inventions

https://history-computer.com/steve-jobs-complete-biography

Read the complete biography of Steve Jobs.

Index